Supplier Improvement Process Handbook

David Sloan
Scott Weiss

AMERICAN SOCIETY FOR QUALITY CONTROL
310 West Wisconsin Avenue
Milwaukee, Wisconsin 53203

i

Supplier Improvement Process Handbook

David Sloan
Scott Weiss

ISBN   0-87389-036-1

iii

Contents

Supplier Improvement Process Handbook

Preface

A company can improve the quality of goods and services purchased from suppliers through the Supplier Improvement Process. The result can be improved quality, price, and service of the company's products, reduced costs, improved profitability, and market share. As summarized in current computer jargon, "garbage in — garbage out," the quality of a product can be no better than the quality of the materials used to produce that product.

Implementation of a supplier improvement process is not simple. It involves changing attitudes and behavior. To be effective, the supplier improvement plan must be integrated into the business plan for the entire organization. It requires patience and persistence, and there is no one right way to do it. Consistent efforts in planning, training, and supplier assistance over a long period of time are the keys to success. The return on investment in time and effort will result in significant quality improvements, lower costs, and better working relationships between the company and its suppliers.

This handbook is not a step-by-step set of "how to" instructions, nor is it intended to be. Rather, it provides suggestions and possibilities, and sets the boundaries for the supplier improvement process to operate effectively. The process is based on real successes that were implemented in a practical manner. Note that the order of implementation does not have to follow the numerical order of the elements.

Introduction

The need for supplier improvement in daily business has been aptly demonstrated within the United States by the impact of high quality and competitively priced products from foreign sources. The effective implementation of a supplier improvement process is essential to the long-term health of any business unit through improvement of market share and profitability.

The system of supplier improvement described in this book was developed and implemented by various 3M divisions in the early 1980s as an extension of the 3M quality improvement process. Lew Lehr, chairman of the board and chief executive officer of 3M in 1981, outlined the company's basic philosophy of supplier relations in a letter sent to the supplier community. He redefined quality as "conformance to requirements" and set a long-term goal for suppliers to provide 100 percent of the company's materials and parts defect free and in line with specifications.

The Supplier Improvement Process consists of four phases: supplier improvement strategy, supplier survey, supplier assistance, and supplier certification/recognition. A sound supplier improvement strategy from upper management is necessary to give direction and support to the process. The supplier survey provides an accurate description of the supplier's quality, engineering, and purchasing/materials control systems. The supplier assistance phase uses survey results to work toward specific objectives with the supplier. Supplier certification entails a reduction or elimination of inspection and testing of the supplied materials, along with appropriate recognition.

Figure 1 shows the Supplier Improvement Process through a variation of the "Control Circle," described by Kaoru Ishikawa in his book, *What is Total Quality Control?* The strategy phase defines the overall goals of the process. A policy or mission statement is appropriate, followed by the formation of a cross-functional management team that develops methods to accomplish the company goals.

Figure 1

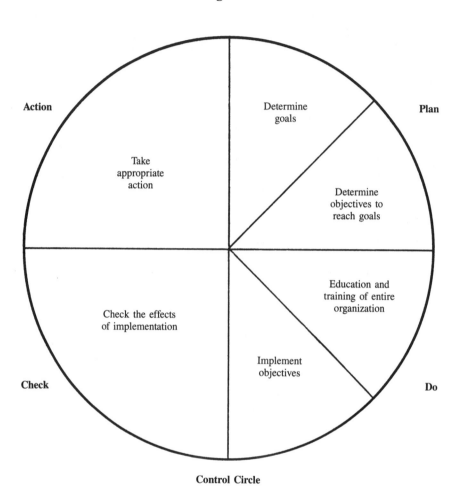

Control Circle

After defining its mission and purpose the team should develop a detailed set of objectives, educate everyone in the organization, and provide appropriate specialized training to accomplish the objectives. Upon implementation of the plan, the team should monitor progress constantly to assure that goals are met, and take appropriate actions.

This handbook outlines the key elements of the four-phase Supplier Improvement Process. Definitions of elements include an explanation of their importance to the overall process. Sample objectives associated with each element are also included where appropriate. Figure 2 shows the Supplier Improvement Process in flow-chart format. The ideas presented here are a compilation of supplier improvement successes generated by many efforts.

Figure 2

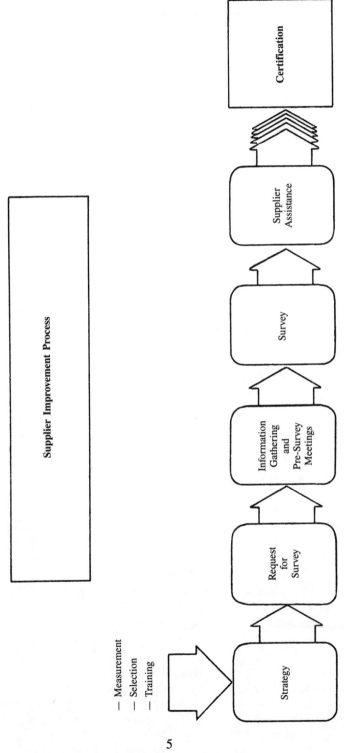

Supplier Improvement Process

Phase 1 — Supplier Improvement Strategy

Management Commitment

The first and most critical element in implementing a successful supplier improvement process is a strong management commitment. Management must accept ownership of the process and integrate supplier improvement into the business strategy of all operating units.

Management needs strong support and involvement of all functions and must develop the goals of the Supplier Improvement Process. Without these goals, the team will not be able to function effectively or meet the expectations of the business. Listed below are examples of types of goals:

- Improve quality of incoming materials (percent conformance to specification).
- Improve supplier/purchaser communication.
- Consolidate supplier base.
- Certify "select" suppliers.
- Facilitate "just-in-time."
- Reduce cost of purchased "A" materials (80 percent of dollar volume).
- Develop 100 percent adequate specifications.
- Reduce cost of quality.

Once goals are established a plan for their achievement follows. The plan is a series of specific objectives, relative to the subsequent elements of the process. For the plan to be useful it must include an objective statement, person responsible, and scheduled date of attainment. A Gantt chart will help the organization to understand the process and lay out the plan. Figure 3 shows a typical supplier improvement plan.

After developing and approving the plan, management should publish and review it on a regular basis. The plan can also be developed into a training manual to serve as a basis for educating the organization about the process and the individual's role in making the process function properly.

Figure 3

18 Steps to Accomplishment	Assigned	J	A	S	O	N	D	J	F	M	A	M	J	J	A	S	O	N	D
1. Obtain Management Commitment	President or General Manager	█																	
2. Form Supplier Quality Improvement Team (SQIT)	President or General Manager		█																
3. Define Purpose & Establish Specific Steps for Accomplishment	SQIT			█															
4. Review & Adopt Supplier Improvement and Certification Guidelines	SQIT				█														
5. Develop & Establish Company Supplier Improvement Guidelines	SQIT						█												
6. Develop & Implement Measurement System for Purchased Materials	SQIT / Plant Team(s)					█													
7. Prioritize & Target Suppliers	SQIT/Plant Team(s)						█												
8. Implement Supplier Corrective Action Process	Plant Team(s)					█													
9. Coordinate Specification Development/Review	SQIT					█													
10. Select & Train Plant Supplier Improvement Teams & Supplier Survey Teams	SQIT								█										
11. Present Supplier Improvement Education/Awareness Program	SQIT						█												
12. Conduct Supplier Surveys	Plant Team(s)							█											
13. Provide Supplier Assistance/Survey Follow-Up	Plant Team(s)								█										
14. Recognition and Certification	SQIT										■	■	■	■	■	■			
15. Follow-Ups (Periodically)	Plant Team(s)																		█
16. Publish Approved Company Supplier Listings	Purchasing							█											
17. Coordinate with Quality Improvement/Just-in-Time Programs and Concepts	SQIT										■	■	■	■	■	■			
18. Document Savings	Plant Team(s)/Cost Accounting											█							

Supplier Improvement Team

A team responsible for the Supplier Improvement Process should be selected to implement the effort. The team may be called a Supplier Quality Improvement Team (SQIT) or any other name that fits the company culture. The key is that top management supports this team and it is made up of decision-making members representing all functions. There should be a formal reporting structure to company management to provide the support and awareness necessary for success.

Members of the team should include personnel from quality, manufacturing, product development/research and development, purchasing, materials control, and plant team leader(s). The team is responsible for developing specific objectives (including who, what, and when) to accomplish company goals. The team also reports progress and accomplishments to management. For large companies with multiple plants, a structure of plant teams reporting to the SQIT is a practical way to implement the process. Plant teams are responsible for administering objectives which include measurement, supplier corrective action, supplier surveys, survey follow-up, supplier assistance, and cost/benefit analysis. For small to medium sized companies, the SQIT and plant teams may be one and the same, or additional personnel may be responsible for specific objectives.

Cost/Benefit Analysis

Like most of the other elements, cost/benefit analysis is tied directly to goals. It is used to measure progress in meeting objectives and to feed back that progress to management. In most cases it is beneficial to assign a dollar figure to the measurement, although in some cases such as improved communications this is very difficult to do. Regardless of objectives, a cost/benefit analysis is needed to sell the process to top management. Some examples of cost/benefit objectives are:

- Reduced cost of quality (appraisal/failure).
- Improved internal operations.
- Supplier consolidation.
- Reduced discrepancies/deviations.
- Improved communications.
- Inventory reduction.
- Lead time, lot size reductions.
- Reduced materials cost.

Consider the following potential cost/benefits of a supplier improvement effort:

1. Lower manufacturing costs will occur due to reduced levels of discrepant incoming materials with subsequent reductions in appraisal (inspection and test) and failure (line stops, scrap, and rework) quality costs. Productivity, yields, and outgoing quality will increase, which will drive down warranty and returned goods costs.
2. A total reduction in the procurement cycle will be realized. The need for "safety stock" will also be eliminated or severely reduced.
3. The Supplier Improvement Process facilitates a true just-in-time system, with *total* confidence in the supplier's quality and delivery performance.

9

4. Improved communications allows for the development of meaningful specifications by becoming aware of each other's needs.
5. By evaluating typical hidden quality costs, we can purchase the true lowest cost, highest quality product.

Specification Development/Early Involvement

This element involves specifications for products currently in production and for new products.

For products in production, it is of utmost importance that the organization improve its specifications to accomplish the objectives of *100 percent adequate specifications*. A key to an adequate specification is one that is mutually agreed on by the supplier and the company. To meet this objective an action plan should be developed. It may be desirable, in a large organization, to appoint someone responsible for this key objective. One obvious task is to define an adequate specification. An adequate specification is one that defines requirements and test procedures. It serves as a basis for accepting or rejecting the material.

Activity should then focus on measuring the percentage of specifications that meet the definition of "adequate" and developing a detailed plan to accomplish the objective of 100 percent adequate specifications. Without the plan in place and resulting improvements the total process is in jeopardy.

For new products, the emphasis should be on involvement of known, high quality suppliers early in the product development cycle. Up-front agreements should be reached such as confidentiality, patent disclosures, and long-term contracts. The benefits of this approach include shorter scale-up time and more manufacturable products of higher quality at a lower total cost. Total cost includes all cost elements — the quoted price as well as scrap, re-work, inventory, line shut downs, and other typically hidden manufacturing costs.

Supplier Measurement

In order to show improvement to management and to suppliers, there must be a system established to measure supplier performance. As an absolute minimum, a measurement of the supplier's product quality and delivery performance should exist. Quality can be measured by percent lots accepted, percent parts accepted, and/or defective parts per million at receiving or in production. Delivery can be measured as percent lots shipped on time, percent lots received on time, and/or number of days early or late. The two key aspects to measurement are use of the same measuring criteria and method at all plant sites, and feedback of performance measurements to the supplier. Examples of the system to measure quality and delivery are shown in Figure 4 and the Appendix.

The Process Capability Index (C_p) is a measurement in use by many companies in conjunction with a specific parts per million quality level requirement to encourage the use of statistical process control. This approach will result in improved consistency by reducing product variability.

It is appropriate to define the requirements for certification while establishing a measurement system. Some key considerations in developing measurements are simplicity, ease of obtaining data, and relevance of the data to the objectives.

Figure 4

Supplier Measurement — Example

	Points for Each Component
I. Quality — maximum 50 points	
1. Percent lots accepted, which includes incoming inspection, line rejects, and reinspections. (50 points for 100% accepted, 49 points for 98% accepted ...)	50
II. Delivery — maximum 30 points	
1. On time — plus or minus one day of due date	30
2. Acceptable — up to and including five days early and two days late	20
3. Unacceptable — more than five days early or more than two days late	0
III. Paperwork — maximum 20 points	
Purchasing and Plant Quality Assurance's assessment of the accuracy and completeness of packing slip, package labeling, invoice, bill of lading, certificate of analysis, and other paperwork.	
1. Paperwork is in good order	20
2. Paperwork needs improvement	10
3. Paperwork is unacceptable	0
IV. Supplier's Attitude and Technical Support — maximum 10 bonus points	
1. Supplier's willingness to cooperate, including changing schedules when necessary	5
2. Supplier's ability to provide changes in procedures or equipment when requested	3
3. Strength of supplier's technical capability	2
	110 points including 10 bonus points

Overall	
Description	**Points**
outstanding	100-110
excellent	95-99
good	90-94
improvement expected	80-89
improvement required	79 and below

The supplier's performance will be rated using a supplier measurement system. This system rates the quality, delivery, and paperwork performance of the supplier on a 100 point scale. Ten bonus points are awarded for positive supplier attitude and technical support. A descriptive rating is used to inform the suppliers of the company's assessment of their performance. The supplier measurement may be used in conjunction with the supplier certification process to track the improvement in the supplier's performance during a specific time interval or for a specific number of lots.

Supplier Selection

Supplier selection has two different meanings. First, it can be the selection of a new supplier for a new product or process, or the replacement of a supplier of a material in production who is not conforming to company requirements. To accomplish this, the survey process should be used. The second meaning of supplier selection, and the intent of this element, is to determine which suppliers to work with in the Supplier Improvement Process to develop long-term partnership relations. The grouping of suppliers in categories narrows the team's focus towards those it intends to work with in a long-term relationship. The suppliers should be grouped by commodities, such as injection molding, machining, stamping, or resins, solvents, papers, etc.

After grouping the suppliers in categories, the team must then consider the organization's goals. To be successful, start with a small number until the process is fully developed. A reasonable number of suppliers for a plant to select initially should be in the range of two to four. Selection criteria can include quality and delivery performance, quality attitude, dollar volume, cost control, inventory/lead time analysis, or criticality of materials (Figure 5). It is critical to communicate the results of the selection process to the entire company, especially the product design or development departments, so that consistency is maintained in the selection of suppliers for new products and early supplier involvement. Specific objectives for this element can include optimizing total number of suppliers, consolidating all purchases with "select" suppliers, identifying candidates for certification, and identifying suppliers to be surveyed.

Figure 5

Supplier Selection — Example

All suppliers are ranked, considering the criticality of the material provided, current quality performance, and annual dollar volume. Points are assigned as follows:

Criticality of Material:	Possible Liability	35
	Possible Line Shutdown	25
	Possible Line Interruption	15
Quality Performance	**Percent Nonconformance**	**Points**
	15+	40
	13-14	35
	11-12	30
	9-10	25
	7-8	20
	5-6	15
	3-4	10
	1-2	5
	0	0
Dollar Volume	"A" Supplier (80% of total $)	25
	"B" Supplier (15% of total $)	20
	"C" Supplier (5% of total $)	15

Training

Training needs for the organization must be identified to support the company's goals. The overall knowledge level of the organization must be taken into account to develop a training plan. For the quality organization formal training can be provided through the American Society for Quality Control or local schools and should include statistical techniques. All quality personnel who deal with suppliers should have appropriate training which may include ASQC certification in quality engineering. All personnel involved in supplier surveys should have survey training. In addition to quality training, the entire organization should be educated as to each individual's role in the process.

Training for the company has been separated from supplier training due to the potential gap in awareness levels in most situations. There will be cases where the supplier can teach the company, or where the knowledge level is the same. In those cases, joint training sessions can be very successful. Examples of objectives for this element include:

- Providing company wide quality training via an outside consultant.
- Training all technical and plant personnel on statistical process control.
- Holding joint company/supplier training courses for just-in-time manufacturing techniques.
- Training a survey team for each location.

Supplier Awareness

Supplier awareness of quality needs is an element requiring direct contact with the supplier. It can vary drastically in activity level, depending on company goals and resource commitment. At the minimum, this could entail a brief presentation to the supplier community to make it aware of your company's quality process. At the other end of the spectrum, it might mean daily or weekly training visits by a group of people to help implement statistical process control. In some cases, a prelude to this element will be a supplier survey (see page 15).

The goal is to initiate the concept of teamwork and partnership. In many instances the survey report will outline the needs of the supplier and serve as the specific action plan.

Using the selection process, the company should first determine the supplier community it plans to work with to meet the goals established. The initial action plan should be developed based on the current level of awareness of both your company and its supplier base. Typical objectives for this element include:

- Holding quality awareness meetings for all suppliers.
- Conducting quality awareness meetings at plant(s) for all select suppliers.
- Surveying all select suppliers.
- Developing communication policy and procedures for all suppliers.

Phase II — Supplier Survey

The supplier survey evaluates a supplier's systems and controls. Its intent is to determine if the suppliers' systems are adequate for the supplier's size and type of business to assure your company consistent quality and service levels. The survey lays the groundwork for an action plan during the supplier assistance phase.

A survey of a particular supplier facility should be performed one time for the entire company. Purchasing should coordinate the surveys and assure that appropriate personnel have the opportunity to participate. To assure thorough and uniform surveys, a survey team consists of at least three functions: quality, purchasing, and engineering (manufacturing and /or laboratory). The survey team(s) should have training to include a review of pre- and post-survey activity as well as the survey process and checklists. In addition, the quality representative should be trained in quality engineering and supplier quality engineering. All surveyors should also participate in training surveys.

There are numerous commercially available survey formats and checklists. *The Procurement Quality Control Handbook*, available from the ASQC, includes evaluative criteria for determining the extent of compliance with ANSI STD Z1.8 *Specification of General Requirements for a Quality Program*. The use of a training process and thorough checklist is strongly recommended, especially if the company has multiple divisions or plants that will be performing this function. Uniform, consistent evaluations are critical.

Once a survey has been completed, the findings are used to define supplier assistance. Typical objectives may include:

- Assuring adequate systems and process capability to allow for certification of specific products.
- Determining and eliminating the system-based causes for frequent, repetitive quality problems.
- Determining whether new business should be placed with a particular supplier.

Phase III — Supplier Assistance

Supplier Corrective Action

The supplier corrective action process is a two-way communication vehicle between supplier and customer. It allows a plant to formally notify the supplier of a discrepant receival. It requires the supplier to take preventive actions by identifying the root cause of the discrepancy and eliminating the cause. It also provides a follow-up mechanism to assure that the corrective action is effective.

The cornerstone of this element is the Supplier Corrective Action Request, a documented form and procedure to assist in improving the quality of incoming materials. The form is shown in Figure 6. The primary concern is that all discrepant materials be analyzed prior to contacting the supplier, since studies have shown that 50 percent of all discrepancies are purchaser caused. It is strongly recommended that each receiving location have personnel responsible for analysis and reduction of discrepancies via the corrective action process. Some organizations have called this person a procurement quality engineer.

SQIT should assure that all plants are using the same process to (1) analyze discrepant material, (2) provide feedback to the supplier, and (3) finalize corrective action via a documented procedure. The use of a flow chart, such as the example in Figure 7, is recommended. All plants should use the same criteria for requesting corrective action and for completing the process.

It is not unusual for an organization to achieve very high levels of improvement, purely by the use of this process. One company reduced its level of discrepant incoming material from 12 percent to 3 percent in a three-year period by using this process, resulting in substantial savings.

Engineering, if responsible for specifications, must be an integral part of this process. The improvement/clarification of specifications is a critical factor in making progress. Without engineering support the process will be stifled. Another key factor in this element is the development and correlation of inspection/test methods, equipment, and results. Specific objectives for this element can include:

- Reducing percent lots discrepant.
- Improving supplier quality and delivery performance.
- Improving specifications for all raw materials.
- Holding regular review meetings with all select suppliers to develop a corrective action plan.

Figure 6
Supplier Corrective
Action Request

Vendor No.		Purchase Order No.	Release No.	Purchasing Agent	Inspection Report No.

To	Attention		From	
	Supplier			
	Address			
	City • State • Zip			

ID or Form Number	Rev.	Lot	Description	Product Engineer
Description of Discrepancy				

Request Date   Supplier To Complete The Following By   Date   And Return To Sender

Cause of Discrepancy and Method of Inspection

Corrective Action Taken To Prevent Recurrence of Discrepancy (Include quality plan)

Date of Implementation	Supplier Signature	Title	Date

Company to complete the following:

Is Corrective Action Satisfactory

Reviewed By ➡   Date

Disposition

Comments	

☐ Corrective Action Acceptable ☐ Corrective Action Not Acceptable ➡ ☐ Survey ☐ Re-source ☐ Re-submit SCAR

Disposition By ➡

Figure 7

Flow Chart

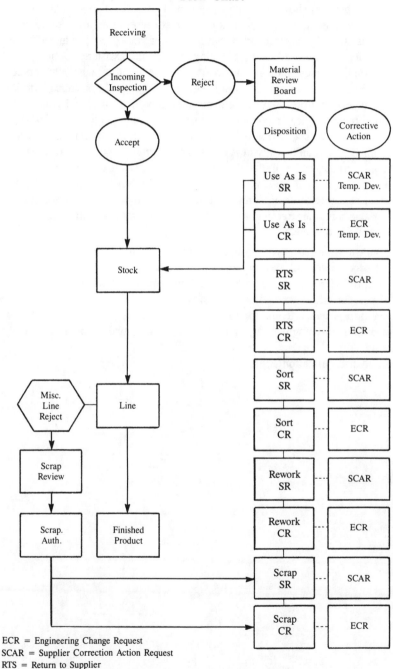

ECR = Engineering Change Request
SCAR = Supplier Correction Action Request
RTS = Return to Supplier
SR = Supplier Responsible
CR = Company Responsible

Supplier Assistance Activity

Supplier assistance activity is the element that can demand the largest resource commitment, but will show the biggest payback if performed well. It allows the company to nurture, strengthen, and expand the concepts of teamwork and partnership. It is strongly recommended that a supplier survey be conducted prior to this activity, and be used as the planning tool for the assistance action plan.

Based on the survey findings, the action plan may require revising the objectives of other elements, especially awareness and training. If assistance is to be a prelude to certification, the two elements should be tied together by mutually supporting objectives. It is recommended that only a very small number of suppliers (two to four) be selected initially for this activity until communications and resource allocations can be fully developed. The learning aspects of this element can be considerable, so be prepared to make major personnel commitments and improvements to internal systems and procedures. The assistance process can be as informal as a weekly visit to review test results and corrective action plans, or as formal as a regular training session on statistical process control. Examples of objectives for this element are:

- Provide assistance to two key suppliers for each plant to implement statistical process control and just-in-time processes.
- Develop, correlate test methods, procedures, and results for all "A" item (80 percent dollar volume) suppliers.
- Develop assistance teams for local, major suppliers.
- Develop quality engineering training program with one supplier for one plant.
- Establish certification action plan.

Phase IV — Supplier Certification

Certification is an understanding between the company and the supplier that the supplier has adequate systems, procedures, and controls in place. It also means the supplier has demonstrated the process capability to assure a consistent quality product delivered on a timely basis. Certification will supplant normal incoming testing and inspection with an audit program, thus relying on the supplier's systems and demonstrated capability to consistently supply products that meet all specifications. It is the cornerstone on which to build a just-in-time process.

Whereas a survey of a supplier's systems is performed once for the entire company by a trained survey team, certification will apply to a specific facility, process, part, or raw material. The company must define its requirements for certification. These should include survey completion, performance feedback system for quality and delivery, adequate supplier systems, demonstrated process capability, demonstrated process controls, and quality, service, and cost objectives met for the specified time period. Minimum quality systems should be consistent with those specified in documents such as ANSI-Z1.8 or meet company-defined requirements.

Formal recognition of certification should be defined during the phase I planning stage. Upon notificaiton, a letter stipulating the parameters of certification should be issued by the company. The letter details the certified products, processes, or facilities, which receiving locations will accept the certified material, and the responsibilities of the supplier and purchaser in the certification agreement. This letter should also define the audit program.

Example objectives of certification include reduce appraisal costs, reduce lot sizes, implement dock-to-stock receivals, and facilitate just-in-time manufacturing.

Recognition

Two forms of recognition defined in this handbook are the Supplier Certification Process and Supplier Councils. Beyond this, there are many forms of recognition that a company can adopt. No matter what the method, the recognition policy should be documented, measurable, and communicated to the entire organization.

One way to recognize good performance is to feed back the supplier's performance measurements on a regular basis. This is so obvious it is often overlooked. Some examples of recognition are:

- Supplier certification (plaque and presentation to all employees).
- Supplier of the year (plaque, letter).
- Performance improvement (X percent better quality than previous period). Letter, group meeting.
- Recognition for all qualified suppliers. Annual supplier day.

The key success factor for this element is to recognize the need and develop a plan at the start of the process.

Supplier Councils

A supplier council is an excellent way to further improve communications and provide recognition to the company's major suppliers. Councils are made up of top suppliers, selected on a commodity-by-commodity basis, along with company management from all key functions. Objectives of the councils are as follows:

1. Formally recognize the company's suppliers.
2. Improve supplier relations.
3. Provide supplier input to new programs and policies.
4. Recommend new ideas to the company.
5. Provide input to the company on solutions to problems with current systems.
6. Provide insight into new and existing supplier initiated programs.

The councils should meet three to four times a year and cover a thorough agenda of topics developed through a brainstorming session conducted at the end of each meeting. The councils can also have task forces made up of company personnel and suppliers to address pertinent topics. Time should be alloted during the meeting for task force meetings and presentations to the members as to accomplishments.

The supplier council concept is an excellent way to improve the communications and business understandings between a company and its major suppliers. A few general guidelines are as follows:

1. All members must be good communicators from top management positions.
2. Include the suppliers on the agenda regularly.
3. Rotate the membership, approximately one third at a time, every year.
4. Choose the members for their individual ability, as well as company responsibility.
5. Do not have competitors on the council together.
6. Have a legal representative present for all meetings.
7. Start preparation for the meetings two to three months in advance.
8. Publish the meeting schedule a year in advance.
9. Name a company manager as council administrator.

Glossary

Audit — An independent review conducted to compare an aspect of performance with a standard for that performance.

Certificate of analysis — Test data from a supplier for a specific lot or quantity of material.

Certificate of conformance — A document signed by an authorized party affirming that the supplier of a product or service has met the requirements of the relevant specifications, contract, or regulation.[1]

Certification — An understanding between purchaser and supplier regarding quality and delivery requirements. The purchaser will eliminate incoming testing and inspection and rely on supplier systems and procedures to deliver the product which meets those specifications. Certification is limited to a specific plant, process, part, or raw material. See Phase IV — Supplier Certification, page 21.

Certified supplier — A supplier who has one or more of his parts, processes, raw materials, or facilities certified.

Control chart — A graphical method for evaluating if a process is in a state of statistical control.[2]

Corrective action — See *SCAR* and Phase III — Supplier Assistance, page 17.

Parameters of certification — Those elements defined by the purchaser that must be met by a specific supplier for a period of time to become certified.

Process capability — The normal behavior of a process when operating in a state of statistical control; the minimum variation achievable after all special causes of variation have been eliminated.

Process capability index (C_p) — A comparison of the process capability (of a process in control) to the specification limits. It is expressed as a numerical value, usually obtained by dividing the tolerance width by the capability (Juran method).

Process capability study — The systematic study of a process by statistical means to discover if it is behaving naturally.

SCAR — Supplier corrective action request. A form and procedure for requesting the supplier to identify the cause of a problem, and to report the action taken to eliminate the cause. See Phase III — Supplier Assistance, page 17.

Select supplier — A supplier who has provided quality products over past years. A select supplier has usually worked closely with the company, and supplies the "A" items that constitute 80 percent of the company's purchased dollar volume.

Ship to stock — A common term to designate a part from a certified supplier with such adequate process capability that incoming inspection is not required.

Specification, adequate — Permits material to be used in a controlled process to produce a product that meets requirements, and serves as a basis for accepting or rejecting the material.

---

1. ANSI/ASQC A3-1978, *Quality Systems Terminology.*
2. ANSI/ASQC A1-1978, *Definitions, Symbols, Formulas and Tables for Control Charts.*

Statistical process control (SPC) — The use of statistical techniques such as control charts to analyze a process for achieving and maintaining a state of statistical control and to improve the process capability.

Subcontractor — A supplier who assumes some or all of the obligations of the original supplier by secondary contract.

Supplier assistance — A period of time when communications intensify between purchaser and supplier. The purchaser and supplier jointly implement an action plan based on information obtained during the supplier survey. See Supplier Assistance Activity, page 20.

Supplier rating — A method of evaluating a supplier's quality and delivery performance and any other important criteria. It is expressed either numerically or comparatively and should be communicated to the supplier as a method of improving performance. See Supplier Measurement, page 10.

Surveyed supplier — A supplier with at least one facility surveyed by a trained three-function survey team. Checklists and survey status reports are completed and turned in. Survey follow-up has been established (action plan) and agreed to by both the purchaser and the supplier. See Phase II — Supplier Survey, page 15.

Survey team — A trained multi-function team that performs a supplier survey. See Phase II — Supplier Survey, page 15.

Survey team leader — The person designated to be leader of the survey team and to coordinate all survey activities.

Test correlation — A measure of the extent to which two tests are related.

Appendix

Supplier Performance by Supplier

Monthly

Supplier Name	Purchase Order Number	Shipment Number	Percent Early	On Time	Percent Late	Quantity Received	Percent Accepted
				Delivery		Quality	
John Doe	123	1	—	100%	—	500	100%
	123	2	—	100%	—	500	100%
	123	3	—	100%	—	500	100%
	123	4	100%	—	—	500	100%
	123	5	—	—	100%	500	95%
	345	1	—	—	100%	1000	95%
	345	2	—	100%	—	1000	100%
	345	3	—	100%	—	1000	100%
Total			12.5%	62.5%	25%	5500	98.6%

Last Six Months Data

John Doe			5.0%	79.9%	15.1%	30,000	99.7%